Good Food

I Like Cereal

By Jennifer Julius

Welcome Books

Children's Press
A Division of Grolier Publishing
New York / London / Hong Kong / Sydney
Danbury, Connecticut

Photo Credits: Cover, pp. 5, 9, 11, 13, 15, 17, 19, 21 by Maura Boruchow; p. 7 © IndexStock Imagery.

Contributing Editor: Jeri Cipriano
Book Design: Nelson Sa

Visit Children's Press on the Internet at:
http://publishing.grolier.com

Library of Congress Cataloging-in-Publication Data

Julius, Jennifer.
  I like cereal / by Jennifer Julius.
    p. cm.—(Good food)
  Includes bibliographical references and index.
  ISBN 0-516-23130-8 (lib. bdg.)—ISBN 0-516-23055-7 (pbk.)
  1. Cereals as food—Juvenile literature. [1. Cereals, Prepared.] I. Title. II. Series.

TX393 .J85 2000
641.6'31—dc21

00-043068

Copyright © 2001 by Rosen Book Works, Inc.
All rights reserved. Published simultaneously in Canada.
Printed in the United States of America.
4 5 6 7 8 9 10 R 05 04

Contents

1	Cereal Shapes	8
2	Cold Cereal	10
3	Hot Cereal	12
4	Cereal with Fruit	18
5	New Words	22
6	To Find Out More	23
7	Index	24
8	About the Author	24

What do you like to eat for breakfast?

I like to eat cereal.

5

Cereal is made from **grains**.

Grains are the seeds of plants.

The seeds can come from corn, wheat, or oats.

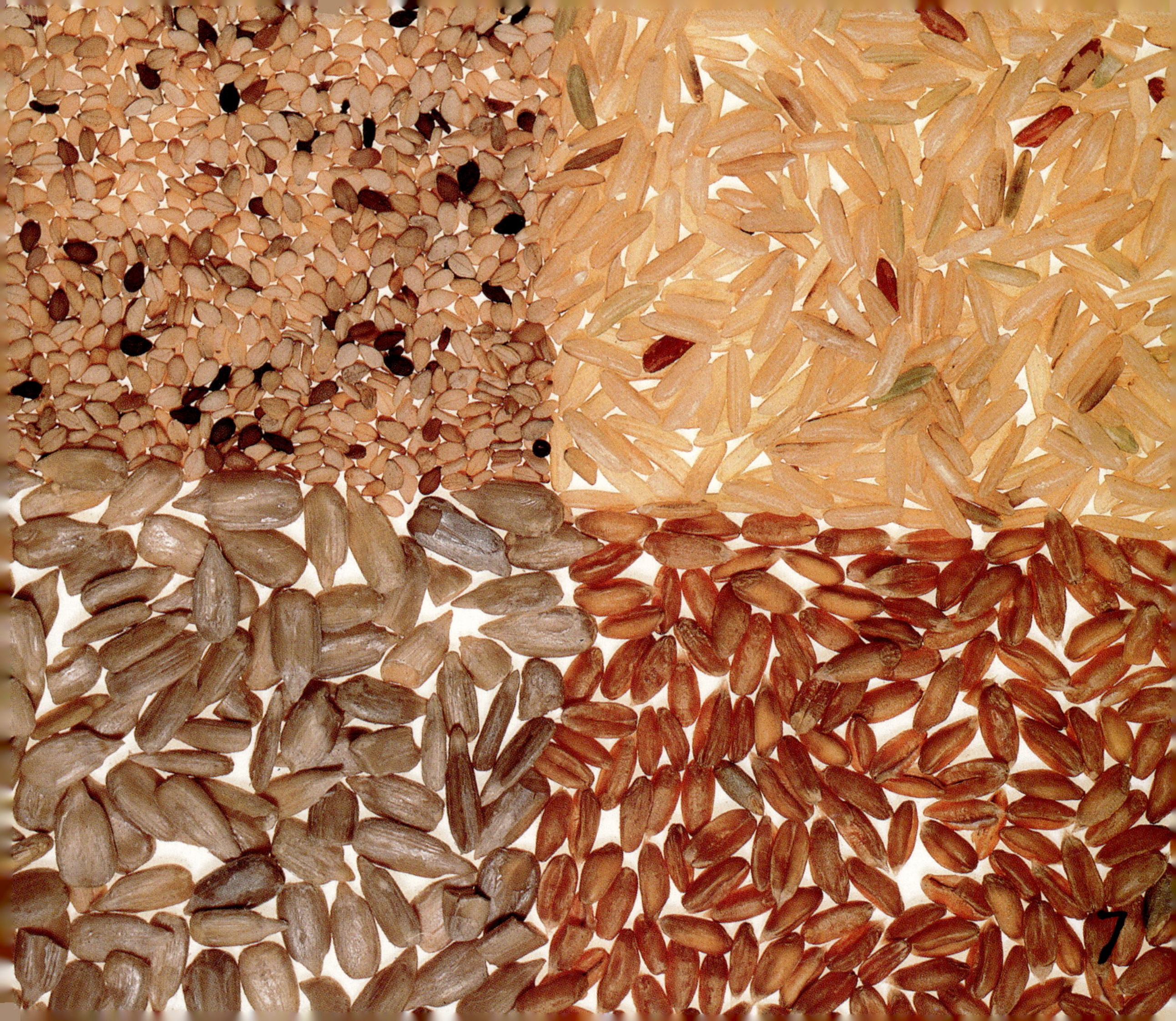

Some cereal is square.

Some cereal is round.

9

Sometimes I eat cold cereal.

I pour cold milk on my cereal.

11

Sometimes I eat hot cereal.

This hot cereal is **oatmeal**.

13

Some cereals have nuts and raisins.

This cereal is **granola**.

Granola is **crunchy** and sweet.

15

Cereal can be many colors.

How many colors can you name?

17

Sometimes I put bananas on my cereal.

I put blueberries on my cereal, too.

19

Which cereal do you like the best?

21

New Words

crunchy (**crun**-chee) something that is hard and makes noise when it is chewed

grains (**graynz**) the seeds of wheat, corn, or oats

granola (greh-**noh**-lah) cereal that has nuts and raisins

oatmeal (**oht**-meel) a hot, cooked cereal made from oats

To Find Out More

Book
My Breakfast
by Heather L Feldman
The Rosen Publishing Group

Web Sites
General Mills—You Rule School
http://www.youruleschool.com/
Play fun games with different kinds of cereal. Play a "make your own breakfast" game and read some silly cereal jokes!

Kellogg's
http://www.nutritioncamp.com/
Find out from Tony the Tiger and Toucan Sam why cereal is healthy. There are lots of fun games and puzzles on this site, too!

Index

bananas, 18

blueberries, 18

crunchy, 14

grains, 6

granola, 14

oatmeal, 12

About the Author
Jennifer Julius is a freelance writer and editor who specializes in educational publishing. She lives in New York City.

Reading Consultants
Kris Flynn, Coordinator, Small School District Literacy, The San Diego County Office of Education

Shelly Forys, Certified Reading Recovery Specialist, W.J. Zahnow Elementary School, Waterloo, IL

Peggy McNamara, Professor, Bank Street College of Education, Reading and Literacy Program